The INSIDE GUIDE

EARTH'S FOUR SPHERES

The Hydrosphere

By Karen McMichael

Cavendish Square
New York

Published in 2023 by Cavendish Square Publishing, LLC
29 E. 21st Street, New York, NY 10010

Website: cavendishsq.com

Disclaimer: Portions of this work were originally authored by Amy Hayes and published as *Earth's Hydrosphere*. All new material this edition authored by Karen McMichael.

Cataloging-in-Publication Data

Names: McMichael, Karen.
Title: The hydrosphere / Karen McMichael.
Description: New York : Cavendish Square, 2023. | Series: The inside guide: Earth's four spheres| Includes glossary and index.
Identifiers: ISBN 9781502664402 (pbk.) | ISBN 9781502664426 (library bound) | ISBN 9781502664419 (6 pack) | ISBN 9781502664433 (ebook)
Subjects: LCSH: Water–Juvenile literature. | Hydrology–Juvenile literature. | Hydrologic cycle–Juvenile literature.
Classification: LCC GB662.3 M385 2023 | DDC 551.48–dc23

Editor: Kate Mikoley
Copyeditor: Abby Young
Designer: Deanna Paternostro

The photographs in this book are used by permission and through the courtesy of: Cover Denis Burdin/Shutterstock.com; p. 4 Joolyann/Shutterstock.com; p. 6 (top) Konstantin Grosch 1986/Shutterstock.com; p. 6 (bottom) Stu Porter/Shutterstock.com; p. 7 Puripat Lertpunyaroj/Shutterstock.com; p. 8 (main) a. v. ley/Shutterstock.com; p. 8 (inset) Patrizio Martorana/Shutterstock.com; p. 9 brgfx/Shutterstock.com; p. 10 robuart/Shutterstock.com; p. 12 (inset) Luisa Leal Photography/Shutterstock.com; pp. 12–13 Katvic/Shutterstock.com; p. 15 I am a Stranger/Shutterstock.com; p. 16 Dario Lo Presti/Shutterstock.com; p. 18 Calin Tatu/Shutterstock.com; p. 19 Kuttelvaserova Stuchelova/Shutterstock.com; pp. 20–21 Sergio Ponomarev/Shutterstock.com; p. 22 Vereshchagin Dmitry/Shutterstock.com; pp. 24–25 Wang Junqi/Shutterstock.com; p. 26 (inset) imageBROKER.com/Shutterstock.com; pp. 26–27 Shandarov Arkadii/Shutterstock.com; p. 28 (top) Gyorgy Barna/Shutterstock.com; p. 28 (bottom) lumen-digital/Shutterstock.com; p. 29 (top) VanderWolf Images/Shutterstock.com; p. 29 (bottom) ND700/Shutterstock.com.

CPSIA compliance information: Batch #CSCSQ23: For further information, contact Cavendish Square Publishing LLC, New York, New York, at 1-877-980-4450.

Printed in the United States of America

CONTENTS

All Earth's systems work together to create balance. Each sphere is different, but they are all interdependent, meaning changes in one can cause changes in the others.

Chapter One

THE WORLD'S WATER

We sometimes take the planet we live on for granted. However, it's a very **unique** place, and if we want to keep it healthy, we need to know how it works. The planet is made of four main systems, called spheres. Each of these spheres has its own processes. However, they depend on each other and are very closely connected.

The atmosphere is the whole mass of air that surrounds Earth. The biosphere is the part of the planet that contains living things. The lithosphere is comprised of the solid outer layers of Earth. The hydrosphere includes all the liquid, solid, and gaseous water on Earth.

In this book, we'll explore the physical parts of the hydrosphere, such as oceans, lakes, and rivers, groundwater, and glaciers. We'll also learn about important processes that involve water.

Fast Fact

Water covers about 71 percent of Earth's surface. It is the most **abundant** natural substance on our planet.

All Kinds of Water

All the water on Earth—solid, liquid, and gas—makes up the hydrosphere. This includes water on Earth's surface, under the ground, and in the air.

A glacier is a large body of ice that moves slowly, either going down a slope or spreading over a big area of land.

Fast Fact

In total, the hydrosphere is made up of 326 million cubic miles (1.4 billion cubic kilometers) of water.

Earth has two types of water: fresh water and salt water. Most of Earth's water—more than 96 percent—is salt water. Less than 4 percent of Earth's water is fresh water. Humans, land animals, and freshwater animals need fresh water to live.

The water in the hydrosphere exists in many different forms. Earth's oceans make up the greatest part of the hydrosphere, while the water in rivers makes up one of the smallest parts of the hydrosphere.

The Start of the Hydrosphere

How did all the water our planet has get here? Scientists have different theories, which are ideas intended to explain

Without the hydrosphere's fresh water, land animals wouldn't be able to survive.

facts or events. Some believe that as Earth was forming, there were many chemical reactions that created water. This water pushed up from the lower layers of Earth's interior to the surface, where it cooled on the surface instead of escaping out of the atmosphere. However, this idea is not as popular as it once was. Many scientists don't think this explains the amount of water on Earth today.

Most of Earth's fresh water exists underground. Sometimes this water comes up to the surface.

A comet is a small, bright object in outer space that contains ice and dust and develops a cloudy tail as it moves near the sun. Long ago, these objects may have left behind lots of water on Earth, creating the hydrosphere.

Most scientists now believe that a lot of the water on Earth's surface actually came from objects from other parts of the solar system. They think that long ago, asteroids and comets that contained ice repeatedly hit Earth, leaving the water behind.

Fast Fact

Degassing is when Earth's interior releases gas, creating water. Earth only gains about 0.07 cubic miles (0.3 cubic km) of water per year through degassing. The amount of water on Earth hasn't changed much since Earth's early history.

Scientists never consider any ideas final because there's always a chance they can learn more about a subject. This scientist is taking a sample of water to study.

THE WATER CYCLE

All of Earth's water plays an important role in the water cycle. During the water cycle, water is circulated between Earth's oceans, atmosphere, and land. This cycle is one of the key processes that take place in Earth's hydrosphere.

There are several steps in the water cycle. Water in the ocean or on land is warmed by the sun's rays. Then, it **evaporates** into vapor that wind carries as a cloud. The vapor **condenses** and falls back to Earth's surface as **precipitation**. There, water may soak into the soil or be added to the water that makes up oceans, lakes, rivers, or streams.

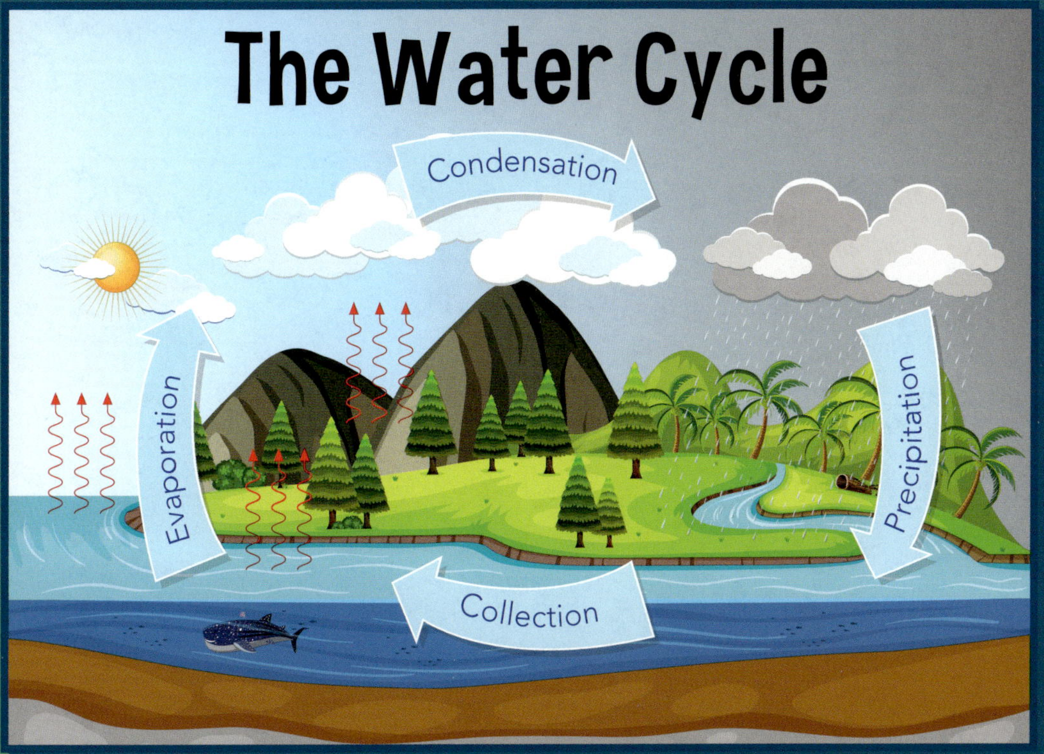

Condensation of the water vapor results in precipitation. When this goes into a river or stream or becomes **runoff**, the water moves back to a lake or ocean, and the cycle continues.

The Southern Ocean is the newest ocean to receive its own name. The boundaries creating this ocean were suggested in 2000. However, not all countries recognize it as its own ocean.

LOTS OF LIQUID

Chapter Two

Most of the water in the hydrosphere at any given time is liquid. The salty water in oceans, seas, and bays makes up most of Earth's water, but there are other important bodies of water too. There's even water flowing in the ground beneath our feet!

The Oceans

Scientists estimate that the world's oceans contain more than 96 percent of Earth's water, which is about 324 million cubic miles (1.35 billion cubic km) of water. This means that oceans are a huge part of Earth's hydrosphere. According to modern-day boundaries, there are five oceans on Earth: the Atlantic, Pacific, Arctic, Indian, and Southern. However, if you look at a map or a globe, you can see that these oceans are all connected, making them one huge body of water. This is sometimes called the Global Ocean.

Fast Fact

Sodium chloride is the scientific term for what we commonly refer to as salt. It's a compound of the elements sodium and chlorine.

Water is made up of two hydrogen molecules and one oxygen molecule. However, ocean water also has some substances that make it different from

The salt you eat comes from deposits left behind by bodies of water!

water found in other parts of the hydrosphere. Sodium, chlorine, magnesium, and sulfate are the four most common substances found in sea water.

Looking at Lakes

Lakes can be found on all of Earth's continents. They come in all sizes, depths, and elevations. Most lakes

Lake Baikal in Siberia, Russia, is the oldest and deepest existing freshwater lake on Earth. At its deepest point, Lake Baikal is 5,315 feet (1,620 meters) deep and holds one-fifth of Earth's fresh water—5,500 cubic miles (23,000 cubic km).

contain fresh water. However, there are some saltwater lakes. All freshwater lakes are open, which means that a body of water flows into or out of them. A lake is closed if water leaves it only by evaporation. Closed lakes become saltwater lakes because evaporation leaves behind salty solids.

Fast Fact

Using photos taken from space, a team of scientists found that there are about 117 million lakes on Earth.

KEEPING EARTH WARM

Oceans play an important role in keeping Earth warm enough for life—including plants, humans, and other animals—to survive. Oceans absorb, or take in, much more heat from the sun than the land or the atmosphere does. As the sun warms the oceans, the oceans keep a lot of that heat instead of bouncing the heat back into space. This helps to keep Earth warm.

Currents move ocean water around Earth, affecting the weather in many places. As currents move around Earth, heat absorbed by the ocean is sent to different parts of the world.

Sometimes rivers create lakes. Runoff, groundwater seepage, glaciers, earthquakes, and volcanic activity can all form lakes too. Lakes depend largely on runoff, so they often get bigger or smaller or hotter or colder due to the surrounding weather conditions.

Water Under Your Feet

Water often flows below your feet—you just don't see it! Groundwater is an important part of Earth's hydrosphere. About 30 percent of liquid fresh water comes from groundwater. Groundwater seeps through spaces between rocks under the soil. A layer of rock or sand that can absorb and hold water is called an aquifer.

Aquifers are key sources of drinking water. Many people dig or drill deep holes in the ground to reach the water stored in an aquifer. These

Depending on where you live, you may use well water. This photo shows a well being created.

Fast Fact

Aquifers are good sources of fresh water, but pumping up too much too quickly can make a well run dry.

deep holes are called wells. A well pulls up fresh water from an aquifer for people to use. Water that isn't pulled up continues to slowly seep through rocks, creating pathways underground. Eventually, these pathways lead to larger water sources such as rivers and lakes. When it comes to Earth's fresh water supply, though, more is found in the ground than in rivers or lakes!

Big, fluffy clouds are called cumulus clouds.

Chapter Three

WATER'S MANY FORMS

If you asked for a glass of water at a restaurant and the server brought you a cup of ice, you might be a little annoyed. Technically, though, they wouldn't have gotten your order wrong. Ice is water frozen from its liquid form into a solid. When it's a gas, water is known as water vapor. So, while most people know that drinking water should be liquid, water in the hydrosphere can be a liquid, gas, or solid!

Water in the Sky

Clouds take on lots of shapes. They sometimes look like balls of cotton. It might be hard to believe, but these fluffy formations are actually made of water, like the rain and the snow that come from them. Up in the air, clouds are part of both the hydrosphere and the atmosphere.

When water is warmed to a certain temperature, it evaporates, or changes from a liquid to a gas. Even though water moves from Earth's surface to the atmosphere as vapor, it remains a part of the hydrosphere. Like the oceans, water vapor traps heat and keeps the air warm. When this vapor travels high into the sky, it sticks to tiny pieces of dust and condenses into tiny droplets. A cloud is made up of millions of these tiny droplets in the sky.

FIND OUT ABOUT FOG

When condensed water vapor is low to the ground, it's called fog. There are several different types of fog. Radiation fog, or ground fog, forms in the evening or at night when the heat from the ground is radiated, or sent out, into the air. This type of fog can stay until morning, when it is said to be "burned off" by the heat from the sun—meaning the condensed water evaporates back into vapor. Precipitation fog is another kind. It forms when rain falls through cold air.

Fog and clouds are very similar, but fog is near the ground and clouds are high in the sky.

Precipitation

When water falls from clouds down to Earth's surface, it's called precipitation. Most precipitation occurs as rain. When small water droplets first form clouds, they aren't heavy enough to fall as precipitation. Over time, they get bigger and heavier, and then they fall to the ground.

Fast Fact

Rain, freezing rain, snow, sleet, and hail are all forms of precipitation.

Temperature is another important reason precipitation occurs. When cold air and warm air

The biosphere, which is made up of the living things on Earth, depends on precipitation as a source of water. For life to exist, there must be liquid water.

meet, it rains. This is because warm air can hold a lot more water vapor than cold air can. When warm air mixes with cold air, the temperature of the vapor lowers, and it condenses, forming more big droplets.

All About Ice

The ice and snow on Earth's surface make up a part of the hydrosphere known as the cryosphere. The cryosphere exists on Earth as glaciers, ice caps, ice sheets, snow, and **permafrost**. The ice found in water, including the frozen parts of oceans, lakes, and rivers, is also part of the cryosphere.

Fast Fact

The word "cryosphere" comes from the Greek word *krios*, which means "cold."

Ice makes up just over 2 percent of Earth's hydrosphere. That may sound like a very small amount, but it's almost 7 million cubic miles (29 million cubic km). Ice that stays frozen all year plays an important part in regulating Earth's temperature. Because ice is so reflective, much of the energy from the sun that hits ice is bounced back into space. This helps keep Earth from getting too hot.

Fast Fact

Glaciers can be found all over the world. They cover about 10 percent of Earth's surface.

About 80 percent of the island of Greenland is covered by a single glacier called the Greenland Ice Sheet.

There are laws and rules against polluting water, but unfortunately, it still happens.

Chapter Four

THREATS AND HELP FOR THE HYDROSPHERE

All of Earth's systems work together to ensure the planet stays healthy. For Earth as a whole to thrive, the hydrosphere needs to thrive too. Unfortunately, some human activities negatively impact the health of the hydrosphere. The good thing is, humans have the power to change the things they do. Together, we can take actions that help, rather than harm, the hydrosphere.

Pollution in the Water

When people pollute water, the hydrosphere is negatively affected. Toxic, or harmful, waste from factories and other harmful substances can leak or be dumped into water supplies and cause plants, animals, and humans to get sick and even die.

Fertilizer is used on farms that grow food and on people's lawns. When it runs off into surface water, fertilizer can cause many problems. One problem is called **eutrophication**. This happens when water has a lot of nitrogen and phosphorous. The excess nitrogen and phosphorous cause plants to grow abnormally fast. This is bad because these plants then use up all the oxygen in the water, leaving behind very little for the other life-forms living in the water.

ACID RAIN

When **fossil fuels** are burned, harmful chemicals are released into the atmosphere. These chemicals can mix with water vapor and change the makeup of water, turning it acidic. When the acid rain falls to Earth, it causes problems. Acid rain can fall anywhere, but it happens most often in places with a lot of factories. However, there are renewable sources of energy, such as solar and wind power, that can be used instead of fossil fuels. If businesses stop using so many fossil fuels and switch to renewable resources, they can help reduce acid rain.

Fast Fact

Forests, lakes, streams, animals, and even people can be hurt by acid rain.

Climate Change

Climate change refers to long-term change to Earth's climate. Most scientists agree that it is caused by human activities. Factories, cars, and large farms release **greenhouse gases** into the atmosphere. These gases stay in the atmosphere and act like the roof of a greenhouse, trapping heat from sunlight in the atmosphere. This causes Earth's surface and atmosphere to have a higher average temperature.

The hydrosphere has been negatively affected by the increase in average global temperature. Much of Earth's sea ice used to stay frozen

Eutrophication can result in harmful algal blooms, or overgrowths of algae.

The melting of sea ice is also harmful to animals, such as polar bears, that rely on sea ice to survive.

all year, but this important part of the cryosphere has been melting more and more, which can raise sea levels around the world. This can cause coastlines to erode faster than normal and increase the rate of floods.

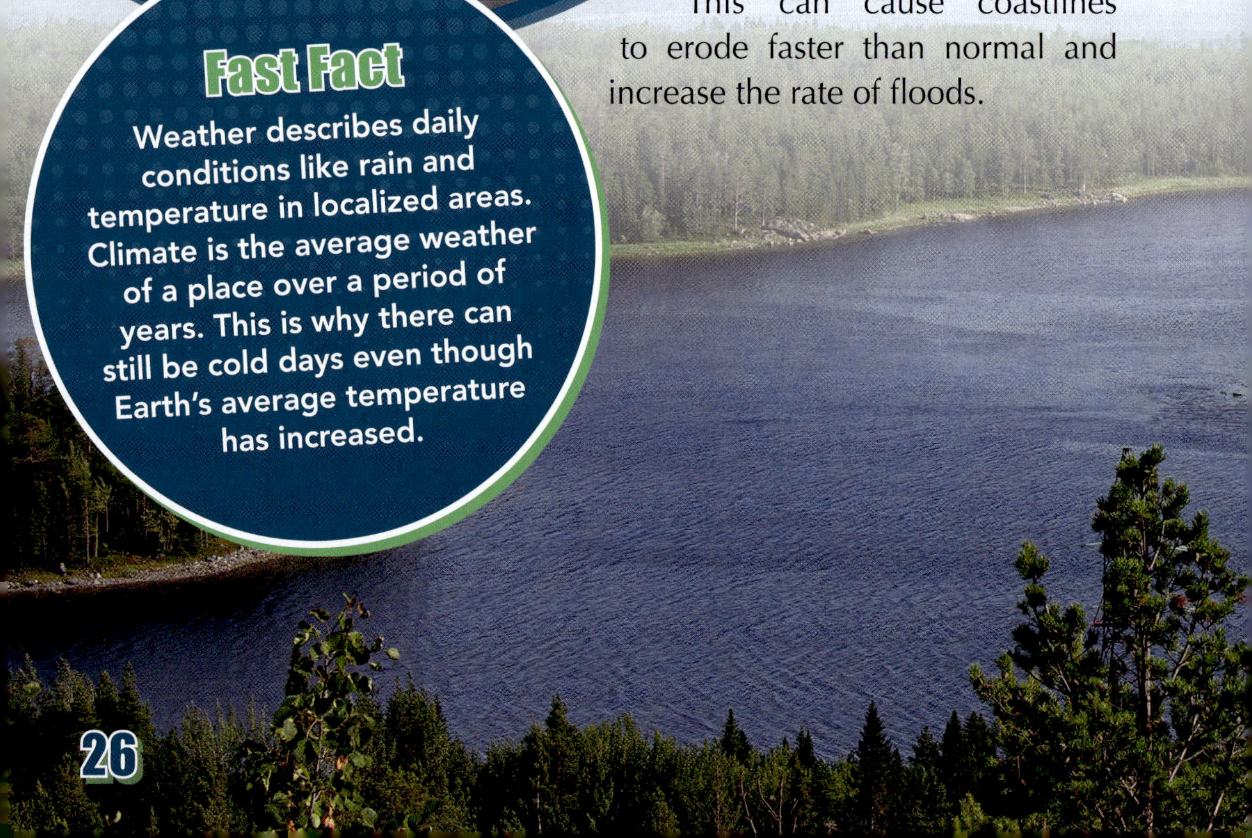

Fast Fact

Weather describes daily conditions like rain and temperature in localized areas. Climate is the average weather of a place over a period of years. This is why there can still be cold days even though Earth's average temperature has increased.

A Valuable Resource

From the water you drink to the water in the air and under your feet, the hydrosphere is all around you. The cryosphere helps keep Earth from getting too hot, while the oceans help keep Earth from getting too cold. Even on a day without a cloud in the sky, the hydrosphere works to keep Earth's weather and climate regulated.

Next time you take a sip of fresh water, think about the journey that water has taken. Spread the word about what you've learned, and tell your friends and family how important it is to protect the hydrosphere!

Fast Fact

Water is Earth's most plentiful natural resource.

With so much water on our planet, it's no surprise the hydrosphere plays such an important role in keeping Earth balanced.

THINK ABOUT IT!

1. Depending on where you live, your drinking water may come from groundwater or surface water. Do you know where your water comes from? Ask an adult to help you search online to learn more about where your drinking water comes from.

2. Think about all the ways Earth's spheres impact each other. What are some ways the hydrosphere affects each of the other spheres (the lithosphere, atmosphere, and biosphere)?

3. What are three things big businesses or governments could do to help protect the hydrosphere?

4. Earth needs water, but water can sometimes also be harmful. What dangers can the different forms of precipitation present to the planet and those who live here?

GLOSSARY

abundant: Existing in a large amount.

condense: To change from a gas to a liquid.

eutrophication: The enrichment of an aquatic ecosystem, such as a lake, with chemical nutrients, which causes an overgrowth of plant life and a decreased oxygen level in the water.

evaporate: To change from a liquid to a gas.

fossil fuel: A fuel—such as coal, oil, or natural gas—that is formed in the earth from dead plants or animals.

greenhouse gas: One of the gases that absorb energy from sunlight, trap heat in the atmosphere, and are part of the greenhouse effect.

permafrost: A layer of soil that is always frozen.

precipitation: Rain, snow, sleet, or hail.

runoff: Water from rain or snow that flows over the surface of the ground into bodies of water, such as lakes, streams, or oceans.

unique: Being the only one of its kind.

FIND OUT MORE

Books

Head, Honor. *Poisoned Oceans*. New York, NY: Gareth Stevens Publishing, 2019.

Sawyer, Ava. *Humans and the Hydrosphere: Protecting Earth's Water Sources*. North Mankato, MN: Capstone Press, 2018.

Smith, Ryan. *Hydrosphere*. New York, NY: Smartbook Media Inc., 2022.

Websites

About the Hydrosphere
mynasadata.larc.nasa.gov/basic-page/about-hydrosphere
Find out more about the hydrosphere and watch a video on the water cycle on this website.

Exploring the Planets: Hydrosphere
airandspace.si.edu/exhibitions/exploring-the-planets/online/solar-system/earth/hydrosphere.cfm
Read more fun facts about Earth's hydrosphere on this website from the Smithsonian National Air and Space Museum.

Hydrosphere
www.nationalgeographic.org/encyclopedia/hydrosphere/
Learn more about this important sphere of Earth on this page.

INDEX